Managing health and safety risk assessments effectively

Managing health and safety risk assessments effectively

James Stowe

YOU NEED THIS BOOK FIRST

London: The Stationery Office

ABOUT THE AUTHOR

James Stowe has many years' experience in the field of health and safety, as a Health and Safety Manager in industry, and as a polytechnic and university lecturer in Safety and Risk Management. He was a founder member and vice-chairman of the National Examination Board in Occupational Safety and Health, and was a long-standing member of the Council of the Institution of Occupational Safety and Health until retirement. He has written many articles on health and safety topics and is currently involved in consultancy work.

31/5/02   00405 PC31549   £17.99

© The Stationery Office 2001

Applications for reproduction should be made in writing to The Stationery Office Limited, St Crispins, Duke Street, Norwich NR3 1PD.

The information contained in this publication is believed to be correct at the time of manufacture. Whilst care has been taken to ensure that the information is accurate, the publisher can accept no responsibility for any errors or omissions or for changes to the details given.

A CIP catalogue recorded for this book is available from the British Library.

A Library of Congress CIP catalogue record has been applied for.

First published 2001

ISBN 0 11 702825 8

Printed in the United Kingdom by The Stationery Office Limited
TJ005515 C20 11/01 657491 19585

Contents

Preface

Ever since the law on health and safety at work introduced the requirement for the employer to identify, assess and control the risks associated with his undertaking, uncertainty as to precisely what is involved has existed. Previous legislation was often precise, prescriptive and absolute; all one had to do was follow instructions. Now, however, compliance is based on decisions stemming from personal assessments of the degree, nature and extent of risks posed by the day-to-day hazards existing in the work environment.

Often, these decisions are subjective and can involve bias and a lack of knowledge or full appreciation of the factors necessary to make an objective appraisal. Risks vary considerably from one concern to another, even when posed by similar hazards. In fact, it is becoming increasingly common for risk assessments to be exchanged between many small- and medium-sized enterprises ('SMEs') to derive an objective level at which to provide the most economical control measures.

Risk assessment and control is now central to *all* health and safety management without exception and it is all too easy to overreact and waste money and other scarce resources on totally unnecessary 'control' measures. In many cases, the actual techniques suggested for carrying out risk assessment are unduly complex, time consuming, inflexible and confusing. Invariably, they leave the assessor with an 'assessment' but fall short of telling him what action should be taken to control the identified risk.

This book takes a good, hard look at *practical* risk assessment and shows how to prevent the materialisation of risks that threaten the assets of the business in question – whatever the nature of the business. It deals primarily with physical risks but not in isolation. Such risks are examined in an holistic context where the social and financial aspects integral with physical risks cannot be ignored. Practical cases of failure to manage risks successfully are provided, along with details of losses incurred and the control measures which, if they had been in place, would have prevented the loss in the first place.

Written for the SME, this book avoids the jargon, matrices, nomograms and formulae so beloved by the risk assessment industry and adopts a practical, down-to-earth, simple and sensible approach to this subject. The prime objective throughout is to improve productivity and profitability by reducing stoppages occasioned by materialised risks and time wasting caused by grappling with unreal assessment techniques.

Jim Stowe
July 2001

Introduction

There is probably more confusion, misunderstanding and downright ignorance about risk
assessment and management than any other
aspect of occupational safety and health. As
the term 'risk' can be defined as 'a threat to an
organisation's assets' amongst other things, it
seems logical to identify, evaluate and control
such risks. If business risks are not adequately
controlled, losses of all kinds can occur and, if
severe enough, can put a company out of
business. If accidents and ill health at work are
frequent events and again, are severe enough,
the consequences can equally threaten a
company's viability. It makes good sense,
therefore, to prevent accidents happening and
ill health developing. Strangely, though, even
the most successful and entrepreneurial SMEs
appear not to understand this simple fact. Consider the following:

'Some years ago, under the heading
"Assessing the risks in a red tape society"
a national newspaper carried an article
about a successful business couple
employing nine people and making
commercial vehicle bodies. The
Management of Health and Safety at Work
Regulations 1992 had just come into force.
The couple stated that these Regulations
required employers to "identify and record
all hazards likely to affect workers or
visitors and show that they have taken

suitable action to prevent accidents". They
considered this requirement to be "a big
burden on time and costs" and that "an
audit of all risks would be a never ending
exercise" and also that "the intention of the
rules (to carry out risk assessments) are
hard to understand".'

Here we have the management of a highly successful SME that has remained viable through
a slump in its market, claiming, in effect, that
it does not understand how to identify and
control its own workplace hazards likely to
affect its own workers or visitors to its premises! Yet that same management has obviously
been involved in much highly effective risk
assessment in managing its business through a
market slump. Its corporate risk management
strategy involving cash flow, market share,
raw materials purchasing, investment and so
on must have been both competent and successful. What, one may ask, is its problem
with identifying and controlling risks stemming from mainly physical hazards in its own
back yard?

No wonder the response of the Health and
Safety Executive ('HSE') refuted the company's complaints stating that risk assessment
and control simply require the application of
common sense. 'People who are doing their
best will not be penalised; they will be given

help and advice', said a spokesman. Incidentally, the same couple were complaining that it would cost them £200 to have all electrical appliances checked 'under another regulation ...'. There is, of course, no regulation requiring such checks to be made.

If one sets one's face *against* risk assessment by adopting a 'head in the sand' attitude, then the assessment process can be a problem. Unfortunately, this is not uncommon in the United Kingdom. We are probably the most reactive of the Western, developed countries; regularly waiting for things to go wrong then wringing our hands and intoning that 'we must learn all the lessons to be learned from this (disaster) to make sure it never happens again'. We never do. And within what seems to be an incredibly short time, a carbon copy incident occurs.

Risk management is about being *proactive*. This means preventing things going wrong in the first place. In every sense this is the preferred course. Expenditure of money and resources on prevention is always far less than paying for reinstatement once the disaster has occurred. The National Health Service ('NHS') is a classic case. If we spent more on the prevention of ill health in the first place, we would not need the vast hospital and medical organisation, which costs billions of pounds annually, to cope with the hordes of sick people we produce in the United Kingdom.

This book takes as its main theme the HSE's assertion that successful risk assessment and management simply requires the application of common sense and the knowledge and understanding of a few everyday terms. It will demonstrate, amongst other things, that

assessing and managing risks is nothing new and that all of us are carrying it out, often without realising it, for most of our waking hours and that it is the basis of every decision we make. It will show the reader how to assess and manage workplace risks by using no-frills techniques which not only work but will prove cost-effective for his business.

CHAPTER 1

The Basics of Hazard, Risk *and* Assessment

No one is certain when *risk* entered the English language. The French version *risqué*, meaning salacious or rude, had been around for a long time but there appeared to be no connection between *risqué* and *risk* as adopted by the insurance industry, for example, to refer to the perils which merchant ships faced when on the high seas. Underwriting risks of many kinds has been the business of insurance companies for generations and insurance actuaries, i.e., those experts who, in calculating premiums, guarantee that the companies win in the end, are some of the world's most competent risk assessors.

THE ROBENS REPORT

In July 1972, the Committee on Safety and Health at Work, chaired by Lord Robens, reported its findings on the health and safety of people at work to the Government. One of the most crucial points in the Report reads as follows:

> 'The primary responsibility for doing something about the present levels of occupational accidents and disease lies with those who create the risks and those who work with them.' (chap.1, p. 7, para. 28)

This statement really set the stage for eventual large-scale self-regulation of industry by those primarily involved in undertakings where risks existed. Thus, risk assessment and control techniques were widespread over a quarter of a century ago and were not ushered in by regulation 3 of the Management of Health and Safety at Work Regulations 1992 as many people seem to think.

RISK MANAGEMENT TERMINOLOGY

When the Control of Substances Hazardous to Health Regulations 1988 came into force, the HSE produced an excellent set of simple leaflets which defined hazard, risk and assessment in the clearest possible way. Although aimed at substances hazardous to health, those definitions can easily be adapted to a universal application:

> 'A *hazard* is a substance, machine, piece of equipment, installation, system of work, article, situation and so on with a *potential* to cause harm, loss or even death.'

> 'The *risk* is the likelihood that the potential *materialises* and the harm, loss or death actually occurs.'

1

People often speak of 'eliminating the hazard'. In some cases this is possible, but very often it is not. Suppose your business is that of wood machining. It is a well-known fact that woodworking machinery is amongst the most dangerous in existence. Therefore, each machine is a hazard with a potential to inflict the most dreadful injuries. If you want to stay in business you cannot 'eliminate' these hazards; you have to reduce the risk of these injuries occurring – by proper guarding, adequate training and supervision, safe systems of work, use of pushsticks, good lighting, good housekeeping, well-maintained machines, and so on.

Risk gives further dimensions to the significance of a *hazard* and has the following two elements:

- the *likelihood* of the hazard materialising, normally on a frequency basis, i.e., once a year, once a day, six monthly, rarely if ever, and so on; and
- the *consequences* of the risk materialising, i.e., catastrophic, negligible, moderate, severe, acceptable, and so on.

Take the example of a bottle of sodium hydroxide solution ('caustic soda') used for cleaning purposes. Locked up and unopened in a cupboard, it is a hazard. It only becomes a risk when out of the cupboard, opened and being used. And the degree of risk will depend on:

- who is using it – are they trained? do they know of its corrosive properties?
- where it is being used – are there other people who may be affected by it nearby?
- how it is being used – splashed around sloppily or careful application by brush?
- the control measures in place – personal protective equipment; spillage control;

cordoning off work area; COSHH data sheet for early treatment of exposure, and so on.

We can see that the definition of risk is less precise than that of hazard and, in practice, determining the *nature* and *seriousness* of a risk in terms of its potential severity and likelihood of materialisation, often present the greatest difficulty during a risk assessment.

DEFINITION OF RISK ASSESSMENT

There is no single definition of risk assessment, but one which is suitable for the SME and accepted by the HSE is as follows:

'Risk assessment is a multi-stage process used to determine the magnitude of a threat (risk) of loss, in its widest sense, to assist management decision making. An assessment should determine whether the risk is tolerable, taking into account existing control measures. If these are not adequate the assessment should recommend more effective measures. Monitoring of these control measures must take place.'

The HSE has published a guide to risk assessment called, simply, *Five Steps to Risk Assessment*, in an attempt to explain the subject to employers and SMEs in particular. We shall be looking at this model in due course. In the meantime, however, it is important to realise that a risk assessment procedure is not rigidly confined to a fixed number of steps. Your own method can include as many, or as few, steps as you decide. The thing is to arrive at a sensible, workable and economically viable decision which will genuinely protect your assets without making you bankrupt.

Risk *and* Decision Making

There are two types of risk – speculative and pure. Speculative risk involves a gamble or a speculation. One makes a decision to gamble on the lottery or the horses or speculate on the stock market. In each case, one's decision will result in one of two possible outcomes – a gain or a loss.

In the case of pure risk, there are again two possible outcomes. Either the risk remains controlled, i.e., it does not materialise, or it gets out of control, or containment, and does materialise, resulting in a loss. In health and safety management we are concerned almost 100 per cent with pure risks. Examples of these are the possibilities of accidents, ill health, damage to property by fire, flood and other natural causes, machinery and plant failure posing danger to employees and so on. Additionally, the SME manager has to manage pure risks relating to theft, fraud, data loss, hacking, sabotage, industrial action, environmental liability, breach of contract, product liability, and so on. The list seems unending.

Some managers believe that insurance takes care of such risks. In the event of *any* of the above risks materialising insurance will only pay a fraction of the costs incurred; uninsured losses are far greater in aggregate than insured ones. And just watch your premiums increase at renewal time! Sensible risk assessment, decision making and overall hazard manage-

ment, then, saves your company money. It is a practical form of insurance and, apart from being a legal requirement from the HSE's point of view *it is very much encouraged by the insurance world.*

Probably the most competent risk assessors in the world are insurance actuaries. These are the people who calculate premiums to insure or underwrite risks of all kinds. They do have masses of statistics amongst other things to assist them in their decision making. But then, the SME manager has sources of information – even his company's accident book – which can help in highlighting trends which will indicate whether certain risks are under control.

In considering health and safety risk assessment decision making it nearly always helps to consider an analogy. For example, if sales of a product or a service are down an investigation into the reason(s) for this is soon mounted and decisions made to counter this adverse development. If accidents, absenteeism or ill health suddenly increase the same investigative and decision making process should spring into action. But both of these examples feature the *reactive* approach. The company should not wait for sales to fall before taking remedial action. It should be constantly monitoring all aspects of the market and the economy to anticipate what

3

CASE STUDY 2.1

```
On 6 January 1968, a 500 ton express train travelling
at 80 mph with 300 passengers aboard collided with a
road transporter carrying a 120 ton electrical trans-
former at an automatic continental type half-barrier
level crossing near Hixon, Staffordshire. Eleven people
were killed and many suffered serious disabling
injuries. The official inquiry into the disaster found
failure to carry out risk assessments on a massive scale
and widespread ignorance on the part of British Rail,
the transporter company, the (then) Ministry of Trans-
port, the police and the owners of the transformer
(then) English Electric.
```

could go wrong. Health and safety risk management is no different. The two key words which should preface every question managers should be constantly asking themselves, are 'What if ... ?'

There were several quite discrete failures in the decision making process which led to the installation of the crossing in case study 2.1, but one merits particular attention.

The Hixon automatic crossing was activated by an oncoming train, in either direction, passing over a switch treadle situated on the track. As the train passed over the treadle, the lowering sequence for the half-barriers was initiated. The familiar flashing lights and bells would operate for eight seconds; the barriers would then commence lowering for another eight seconds and after a further eight seconds the train would arrive at full speed at the crossing. So from the commencement of the ringing bells and flashing lights, anything or anyone on the crossing had 24 seconds to get

clear and nothing on earth could stop the train arriving at the prescribed time.

The transporter was stuck on the crossing. Even if it had been able to move its length and maximum speed (6 mph) would not have enabled it to clear the crossing in 24 seconds. But British Rail had posted a notice at the crossing stating that 'Vehicles must not become immobilised on its crossings' and quoting a relevant bye-law.

In the case of this particular failure (one amongst many) no one had asked the question 'What if ... something *does* become immobilised on the crossing?' which, in retrospect, seems incredible. But one of the first lessons a SME manager can derive from this case study is the risk posed by the general lack of respect accorded to notices with a *negative* message, e.g., 'no smoking', 'no entry', 'authorised personnel only', 'keep this space clear', and so on. So if your risk control measures are dependent on members of your workforce either

CASE STUDY 2.2

At a processing factory, the majority of employees are bussed to and from the factory. Although there is always spare seating capacity on the buses, there is a mad stampede at finishing time to get the front seats on the buses so that the workers can be 'first off' when the buses reach the drop-off points. This saves those first to alight at least 40 seconds.

The corridor to the factory exit has a concrete floor with notices ordering people not to run along the same. It is studiously ignored and when a worker stumbles and breaks her ankle the union solicitors claim compensation on the grounds that the company is in breach of section 2(2)(e) of the Health and Safety at Work, etc., Act 1974 (safe access and egress). Threats to discipline workers who are caught running are likely to precipitate industrial action.

doing something specific or, alternatively, *not* doing something that is required of them you should ask yourself 'What if ... ?' they fail to do what they are instructed to do or ignore instructions to refrain from adopting certain patterns of behaviour.

It must be appreciated that the risks at work often stem from situations similar to the one outlined in case study 2.2. Risks often arise from the way people behave at work rather than the equipment with which they work or the processes, chemicals and machinery involved. In the case of the above risk, two extra exits were installed close to the existing one so that there was no need to run the 50 yards or so to the original exit. The two extra exits allowed all workers to reach the outside of the factory very quickly without any congestion developing. Tripping and stumbling hazards were, in this case, virtually eliminated.

COST BENEFIT ANALYSIS

Decision making in its simplest form consists of:

- defining a problem or situation;
- evaluating the options available to solve the problem or improve the situation;
- implementing the preferred option(s); and
- monitoring the progress post implementation.

During the evaluative stage, cost benefit analysis is almost always used. Cost benefit analysis decides whether the benefits likely to flow from a course of action where a cost has been incurred are greater than that cost. In the above processing factory example, the cost of the bussing programme was outweighed by the benefits of good punctuality, lower absenteeism, retention of employees on site at lunchtime (the factory was in a greenfield site) and no drink/drugs problems. Also, the cost of providing the two extra exits was more than justified by the consequent elimination of the running problem with its attendant tripping/stumbling risks. The compensation paid to the lady who suffered the broken ankle nearly amounted to the cost of the new exits.

It is perfectly proper to apply cost benefit analysis to health and safety risk assessment decision making. In fact, the Health and Safety at Work, etc., Act 1974 uses the term *reasonably practicable* to describe the legally acceptable relationship between a risk and the cost (in terms of money, resources and effort) which would be considered appropriate to control that risk. In other words, it is perfectly acceptable, in law, to consider the costs of controlling an identified risk and if the risk is insignificant in relation to the costs it is **not** reasonably practicable, i.e., necessary, to spend money, etc., on control measures.

CASE STUDY 2.3

For decades, windows on railway carriage doors could be opened fully allowing a vertical opening of some two-and-a-half feet. This permitted, amongst other things, good ventilation on crowded commuter trains in hot weather. One day, a person put his head out of an open window at the wrong time. As a result, the railway company concerned modified every window so that it could only be opened a few inches thus creating conditions of acute discomfort for thousands of its customers.

CASE STUDY 2.4

A company spent thousands of pounds providing extra external earth connections to fixed electrical equipment much of which was installed on wooden boards inside brick buildings. All the equipment was already earthed internally via the cabling through which it operated. The company had been advised that there was a 'risk of electric shock' from the metal casings of the switchgear, etc., and that the earthing was recommended by the Electricity at Work Regulations 1989. The contractor concerned no longer has the standing electrical contract with the company.

Some decisions as to what constitutes an unacceptable risk are based on ignorance.

Leaving aside the ethics of the above incidents, they are good examples of excessive expenditure on control measures for risks which, in real terms, did not warrant this.

ACCEPTABLE RISK

From the foregoing it will be seen that there *is* such a thing as 'acceptable risk'. This is also known as tolerable or residual risk and the concept is recognised by the HSE. It can be defined as the level of risk considered to be acceptable by an undertaking. It also means that the risk has been reduced to the lowest level which is reasonably practicable. A selection of factors which go towards determining acceptable risk include:

- the costs of further reducing the risk;
- the benefits of existing control measures; and
- the probability (high, medium, low) of materialisation of the risk, together with its consequences.

CASE STUDY 2.5

The housing officer of a city council decided that window boxes belonging to tenants of the council's flats were a hazard and posed a risk of falling onto the heads of passers-by. The same council previously decided to cut down mature horse chestnut trees lining the council's streets because it considered they too posed a hazard. This time it was considered that there were risks of:

· children being run over as they collected conkers;
· motorists skidding on squashed conker husks; and
· people being felled by sticks thrown by children to dislodge conkers.

The same council will not permit bouncy castles to be erected on its land.

The first two decisions are being rescinded but the bouncy castle ban stays.

CASE STUDY 2.6

A college science lecturer performed a routine curriculum-based experiment without first purging the apparatus as was required as a routine step. He also failed to issue the mandatory face protectors to the students against standing orders. There was an explosion and some students sustained slight temporary skin discomfort from the explosion.

If the rules had been followed, as on numerous occasions in the past, the experiment would have been perfectly safe. As it was, the Local Education Safety Officer, who came to hear about it, had it removed from the curriculum as being too great a risk.

CASE STUDY 2.7

```
Long-term repairs to a railway bridge which carried a
busy commuter road over a cutting caused considerable
delays due to only one lane being open at one time. An
ex-soldier  from  the  (then)  Royal  Electrical  and
Mechanical Engineers ('REME') wrote to the local paper
suggesting the erection of a Bailey Bridge alongside
the existing railway bridge to ease traffic congestion.
The ex-REME soldier had surveyed the terrain and
concluded it was ideal for a Bailey Bridge.

The County Highways Engineer replied that he could not
risk the life and limb of the public by erecting one of
these temporary structures which, 'after being used
once', were discarded as 'too dangerous to be used
again'. After he had been educated in the legendary life
and reliability of these structures, one was duly
erected alongside the railway bridge in question.
```

The last of the above bullet points is one that should be very carefully considered because *some* perceived risks can border on the ridiculous.

Some decisions as to what constitutes an unacceptable risk are based on ignorance.

The purpose of case studies 2.3 to 2.7 above is to emphasise the importance of having risks assessed and decisions made on control measures that are realistic, sensible, economic and cost effective. In all the above cases, bad decisions were made in the first instance. These were changed by public demand as much as anything but go to show how easy poor decisions, based on unrealistic assessments of 'risks', can so easily be made by people not competent to make such decisions. It has to be remembered that there is no such thing as a risk free society.

VOLUNTARY AND INVOLUNTARY RISK

Managers should be aware that employees are often prepared to take much greater risks when not at work than when at work. For example, people will often indulge in high risk sporting activities – using the correct safety gear, of course – without a second thought about their safety. This is because they see themselves as being 'in charge' of the situation and they can withdraw if they feel like it. At work, however, they do not see themselves as being 'in charge' to the same extent and are thus reluctant to accept the same level of risk.

On the other hand, workers will often put themselves at risk to prevent something going wrong at work. For example, there are well-documented examples where workers have been injured by attempting to correct the feed of products to strapping or shrink wrap machines; to prevent items falling off conveyor belts and to sustain back injuries through struggling with overloaded trolleys with badly maintained castor wheel arrangements. Human behaviour at work when confronted by risk is dealt with in chapter 5, below.

CONCLUSIONS

The topics and case study examples in this chapter are intended to inject a sense of realism into the subject of risk management. In many instances, in both the private and the public sectors, massive inconvenience is caused to many people by situations created 'for safety reasons' by people in authority who have decided that an unacceptable risk exists. Often, these situations are really for the convenience of those creating them as it is quite difficult to see precisely what risks exist.

However, this 'safety at all costs' mentality is being fuelled, unfortunately, by the growing compensation culture in the United Kingdom. This worrying development has not yet been officially linked with risk assessment programmes but it no doubt will be in the not too distant future.

What practical course of action can an employer take to protect himself – as far as is practicable – from the likelihood of such compensation claims? Risk assessments based on the general and approved industrial practice is the primary guide to the standard of care the employer has to provide. These approved standards are widely published by the HSE in a variety of forms – leaflets, booklets, Guidance Notes, Approved Codes of Practice, bulletins, and so on. Bear in mind that none of these standards place intolerable loads on SMEs and generally reflect the best, safest and economical means of carrying out a job.

Remember that all you need to do to get on the information trail of *any* article, substance, process, work activity, etc., to enable a practical risk assessment to be made, if you are unsure about how to proceed is to ring the HSE Info-line on 0870 154 5500. You will be told what literature is available – much of it will probably be free – and you can order it from HSE books there and then. Much cheaper than calling in a consultant!

CHAPTER 3

Risk Assessment in Practice

A risk assessment is a careful examination of what, in your work and workplace, could cause harm to people (employees, the public, visitors, contractors, and so on). These sources of harm are the *hazards*. Once you have identified these hazards you have to decide whether your existing precautions to control or contain them are sufficient to reduce the *risks* they pose to an acceptable level and are adequate to prevent harm. Assessing risks is a legal requirement but the assessment is only part of the process. Control measures have to be implemented and monitored, records maintained, and reassessments have to be carried out when there are any material changes in the nature of the hazard. For example, the Health and Safety (Display Screen Equipment) Regulations 1992 require risk assessments to be made of all workstations where display screen equipment ('DSE') is used. If the workstation is moved or different equipment, furniture, lighting, ventilation, etc., is provided then this constitutes a material change and the workstation risks must be reassessed.

EXAMPLES OF WORKPLACE HAZARDS

These cover an extremely wide range, from the obvious such as a damaged electrical wall socket, defective work equipment or torn floor covering creating a tripping hazard, to the more obscure such as a faulty heater producing carbon monoxide or short cuts/unsafe work practices involving unsupervised/inadequately trained employees.

CASE STUDY 3.1

A temporary laboratory assistant recruited from an agency was preparing mineral samples for testing by boiling them in perchloric acid. A beaker of hot acid slipped from the assistant's gloved hands and spilt over her cotton laboratory coat.

Perchloric acid is unstable and is considered a fire hazard when mixed with carbonaceous materials. The assistant's laboratory coat ignited spontaneously with the hot acid causing severe burns to her arms, legs and body.

While examples of the obvious should be easily identified and can normally be noted by a walk round inspection with a clipboard, the less obvious hazards are another matter.

The system of work involved in case study 3.1 is one which requires very careful risk assessment in view of the number of hazards involved. They include:

- Perchloric acid – unstable, very toxic, dangerous to respiratory tract, mucous membranes and skin. Can ignite spontaneously when in contact with cotton and other carbonaceous materials.
- Hydrofluoric acid (also used in above laboratory) – very toxic, produces severe skin and tissue burns which are very slow in healing, emits highly corrosive vapours when heated.
- A hotplate for heating beakers of acid.

A proper risk assessment should have elicited the following facts, amongst others:

- The assistant was a temporary employee from an agency and may not have been 'up to speed' on the sample preparation process.
- She should have been wearing a full frontal impervious apron over her cotton laboratory coat.
- Goggles or full face protection should have been provided rather than the standard safety spectacles she was wearing at the time.
- The dangerous properties of the acids in question should have been instilled into her by sufficient and adequate training and initial supervision.
- There would be a need to regularly check the effectiveness of the extraction system in view of corrosive vapours given off by the heated acids.

The less obvious hazard might have been the fact that while acids, and in particular, heated acids are dangerous enough the instability and spontaneous fire reaction property of perchloric acid might not have been known to the assistant.

TWO MAJOR FAILURES IN RISK ASSESSMENT

(1) The Hixon level crossing disaster

The disaster outlined in case study 2.1 includes several additional hazards to the 24 second barrier lowering sequence which were not identified when the whole crossing concept was approved by British Rail ('BR'). They include:

- The fact that the change from the traditional British, sturdy, full gate crossing, operated by a signalman or crossing keeper to the flimsier, automatic half-barrier device represented a *lowering* of safety standards for the United Kingdom was not appreciated by BR.
- The notice by the telephone for drivers of heavy or long loads to contact the distant signalman for permission to cross was worded '**IN EMERGENCY** telephone signalman for permission to cross'. There was no emergency when the transporter started to cross. The wording of these telephone notices was changed after the disaster.
- The transporter had the usual police escort. Neither the transporter crew nor the police escort were clear about the police responsibilities. (These are to warn other road users of the presence of a large load on the road – nothing else.)
- The route plan for the transporter from the English Electric factory to the disused airfield hangar used for storage was

11

approved by the (then) Ministry of Transport 'Bridges Section'. This section was only concerned with the strength of any road bridges the transporter might have to cross.

- The haulage firm in question had experienced a near mishap at another continental crossing some weeks before Hixon. The haulage crew in question reported to their head office 'that the train arrived extremely quickly after the barriers had lowered' and that they had just cleared the crossing in time. The head office did not circulate this information to its national depot network.

So with all the various authorities, utilities, public bodies, etc., involved, *not one of them* foresaw any of the hazards set up by the conversion to continental half-barrier crossings and the enormous attendant risks stemming from these hazards – until it was too late.

(2) The explosion at the Nypro (UK) Ltd works

The explosion at the Nypro (UK) Ltd works at Flixborough was due to the ignition of a massive cloud of cyclohexane vapour which had escaped from a badly engineered pipe connecting two reactors. The pipe replaced one reactor which had developed a crack in its outer mild steel skin and which had been removed for examination. There were several failures to assess risks which led to the devastating explosion – reported to be the biggest in Britain since the Second World War. They include:

- The plant management restarted production without ascertaining the reason for the crack in the reactor skin and without examining the remaining five reactors for similar cracks.

- There was no mechanical engineering expertise on site of the level of competence required to design, test and install a proper bypass pipe.
- The management at the time consisted of chemical engineers without any mechanical engineering knowledge. They did not realise how dangerous it was to run the plant without adequate mechanical engineering expertise.
- The company safety officer advised against restarting the production process without ascertaining the reason for the cracked reactor but was overruled.

These two major disasters illustrate how multi-faceted risk assessment often is and while a SME manager is unlikely to face hazards of a similar magnitude, it is worth being aware of this feature. All too often, a physical hazard is identified, considered to be a potential source of harm (a risk) and subjected to some form of control, repair, guarding or similar and then regarded as fully assessed. This is often only part of the risk assessment process.

A SIMPLE MODEL FOR ACCURATE RISK ASSESSMENT

All the examples of inadequate assessment and control of risks mentioned so far have a common theme running through them. This is the *human* failure. And there is a truism – often denied by those accountable for failure to prevent risks materialising – which states:

'*All* failures are ultimately *human* failures.'

When disaster strikes, the speed with which everyone appears to absolve themselves of accountability or blame makes one wonder whether the incident actually happened as nobody seems responsible! So one of the early stages of risk assessment is to make people *accountable* for assessing and controlling risks in their own domains and accepting responsibility when things go wrong. One of the best known authors of safety information, Dan Petersen, states in his book *Techniques of Safety Management* that 'managers who are held accountable will perform' – a very true statement.

If we accept that 'all failures are ultimately human failures' – and we really have no option but to accept this – then the human factor must feature quite prominently in *any* risk assessment exercise, whether this is in the form of a major hazard analysis or a modest assessment exercise at a SME. The model being proposed is shown diagrammatically below:

PLACE
(e.g., office, workshop, factory, laboratory, fabrication shop)

Level of
RISK

PEOPLE ←——→ **METHOD**
(e.g., workers, visitors, members of the public)   *(e.g., worksystem, shortcuts, production pressures, perception, characteristics, skills, experience)*

It will be noted that the level of risk assessed as represented by a particular situation is the resultant of three factors; the actual work environment, the people in it and the patterns of behaviour being adopted by those people.

DERIVING THE LEVEL OF RISK INVOLVED

Consider a work environment comprising a busy, fairly crowded metal fabrication shop before start of work in the early hours of the morning. The shop is empty and the machinery is switched off and silent. The workplace probably bristles with hazards but at that time poses no risk because no one is present to be injured or otherwise affected (rather like the bottle of sodium hydroxide mentioned in chapter 1; locked up in a cupboard it is a hazard; it only becomes a risk when it is opened).

If we now introduce the workforce into the above metal fabrication shop, the level of risk has risen slightly due to the presence of people and their *proximity* to the shop's hazards. However, at this moment, they are simply standing motionless so the risk of injury is slight.

Once work commences and the employees start to interface with the machinery, plant and equipment, the risk level can, and usually does, increase dramatically. So the best time to assess the risks from hazards in this fabrication shop is when the workers are actually performing their allotted tasks and not when the shop is empty. It might be argued that the hazards are still observable whether the employees are present or not and this is true. However, as has been stated already, some hazards can result from the way people work and it is, therefore, essential that these are noted along with the traditional physical ones.

THE DYNAMIC ASPECTS OF RISK

Further consideration of the above model will reveal that none of the three factors – Place, People and Method – is necessarily static; all are, in fact dynamic in that they can change quite rapidly. The factor *least* likely to change frequently is the place, i.e., the office, factory, workshop, etc., where people are actually at work although some workplaces can often be far from static.

The people in the workplace can be subject to change quite frequently. People leave, are off work temporarily, agency staff and temporary workers come and go, so there is definitely a dynamic aspect about the people factor.

Thirdly, the method or system of work changes more frequently than managers realise. It is said that there are three ways a job can be done:

- the way management say it should be done;
- the way management think it is being done; and
- the way the workers are actually doing it.

When a presumed 'safe' system of work has been designed as part of a risk control pro-gramme and is being departed from for, e.g., reasons of 'production pressures', threatened penalty clauses or threats to bonus earnings, the whole risk assessment relating to that job is compromised. One can do worse than heed the motto of the chief projects engineer of a successful American company with an exemplary safety record:

'My aim in life is to get the job done on time, on budget and safely.'

So risk assessments must be based on empirical, real world situations which reflect how a job is actually done and not on the nor-mative, i.e., how a job *should* be done.

CASE STUDY 3.2

A happily married employee, generally quite content with his lot, works with equipment which requires a high degree of concentration to avoid accidents which can cause severe injuries. He finishes work on Friday, look-ing forward to a leisurely weekend, expecting to return to work on Monday for another uneventful week's work. Instead, he is greeted with an unforeseen financial cri-sis, has a violent quarrel with his wife and is visited by the police who inform him his teenage son has been caught dealing in drugs. Within half an hour of starting work on Monday, his concentration lapses to such an extent that he has a serious accident with his equipment. The enforcement officer investigating the accident (reportable under RIDDOR) asks to see the risk assess-ment and training records for the employee concerned.

OTHER 'PEOPLE' FACTORS

There are a number of other people-orientated factors which contribute to the variable risk level of an assessed workplace and which relate very closely to the method of work factor in the above model. Most are, to a great extent, related to stress which is now recognised as a hazard at work which creates very real risk.

The system of work relating to the scenario described in case study 3.2 would have been assessed prior to the accident with the operator working reliably, with full concentration, carrying out the correct sequences and being aware of the risks involved in the job. A level of risk would have been assigned, with any risk control measures deemed appropriate in place. Work would have proceeded without any untoward event.

On the Monday a significant change in the level of risk would have occurred but unless a reassessment of this was made by, say, a supervisor who was aware of the employee's distressed state of mind, the increase in the risk level would probably have gone unnoticed. In actual fact, this is what happened and the accident occurred.

APPLYING THE HSE'S FIVE STEPS TO RISK ASSESSMENT

The HSE's booklet *Five Steps to Risk Assessment* can easily be applied to the above case study as follows – in this case using the first three steps:

- *Step 1 – Look for the hazards.*

The hazards are already well documented; the job in question involves equipment which is hazardous in its own right and requires concentration, competence and experience to operate without undue risk (an item of woodworking machinery is a good example). If these attributes are all present and the equipment is properly maintained and adjusted, then the risk might be said to have been reduced to the lowest level which is reasonably practicable.

- *Step 2 – Decide who might be harmed and how.*

In the case of the woodworking machine, the most likely person who might be harmed is the operator. Where other plant, machinery or equipment is concerned, other people in the vicinity could be affected.

- *Step 3 – Evaluate the risks (of someone being harmed) and decide whether the existing precautions are adequate or whether more should be done.*

The initial risk evaluation showed that the risk level was low. The operator was the only person in proximity to the machine, he was trained, experienced, competent and reliable with a normally untroubled, placid disposition. The 'existing precautions' or control measures, therefore, could justifiably be considered adequate.

All this changed on the fateful Monday when the operator was under considerable stress as events showed. Had management been aware of the operator's state of mind, a sensible control measure would have been to transfer him temporarily to less risky duties.

STRESS AT WORK – MANAGEMENT'S RESPONSIBILITIES

Stress at work, of course, is a known hazard which is capable of causing serious risks – not only to the worker suffering from stress, but also to colleagues dependent on the sufferer for their own safety. An employer is duty bound to assess the risks where stress is caused or made worse by work. Where stress is caused by problems *outside* work, the employer is expected to show understanding when non-work problems make it difficult to cope with the normal pressures of work. But where workplace risks are created or existing ones exacerbated the employer must make the necessary reassessments of risks as demonstrated by case study 3.2.

The validity of the above model can be amply demonstrated by considering how the risk level would rise if an experienced, mature and dependable workforce was replaced by a young, immature, untrained and irresponsible one (*People*). Or if new equipment or plant were to be installed and commissioned without first training the operators (*Place*). The changing of a system of work without first consulting and discussing with the employees expected to operate the new system can also create new risks (*Method*).

AN ANALOGY

If thinking about risk assessment in terms of 'three dimensional dynamics' sounds both daunting and burdensome, think of your attention span when driving a car. You are, in fact, carrying out the very same risk assessment the whole time you are mobile. You are negotiating a *place*, i.e., the highway, for the entire duration of your journey. It is constantly changing. You are assessing both the behaviour and the likely behaviour of *people*, e.g., pedestrians, cyclists, motor cyclists and other drivers, which is also constantly changing. Finally, you adopt a *method* of driving which is not only legal but ensures, so far as is practicable, that you reach your journey's end safely. This is changing by the second. A good driver, therefore, would appear to be a good risk assessor.

Many requirements under safety and health legislation have common sense parallels in everyday life but because of the barrier put up to health and safety at work by so many employers many of these parallels remain hidden. People are subconsciously assessing risks and making decisions many times a day but do not actively think of the process as such.

CONCLUSIONS

The primary aim of this chapter has been to stress the need to look further than the inanimate, physical, chemical, biological, etc., hazard itself when carrying out risk assessments. It is vital to take into account the people involved or interfacing with the hazard and the way they are working with it when determining the level of risk applying. It has been demonstrated that the same hazard can pose differing degrees of risk depending on the competence of the relevant employee(s) and the way they behave when interacting with it. Differing degrees or levels of risk mean *higher or lower costs when providing controls*. As a young trainee machine operator proceeds to gain competence and experience through being provided with information, instruction, training and supervision the cost of this provision in terms of time and money will progressively diminish as the risk reduces.

CHAPTER 4

The Assessment of Hazards

Probably the first large-scale risk assessments in industry took place when the Control of Substances Hazardous to Health ('COSHH') Regulations 1988 were introduced. These Regulations required employers to assess the probable levels of exposure of employees to all manner of substances – liquids, solids, gases, vapours, fumes, and so on – which were hazardous to employees' health. The risk assessment process was really no different then to the recommended process today as outlined by the HSE's *Five Steps to Risk Assessment*. The COSHH situation is probably easier because the assessment in the main deals with specific identifiable commodities with information on their hazards provided by manufacturers or suppliers.

MISUNDERSTANDINGS

Nevertheless, misunderstandings – many resulting in much unnecessary expense on so-called 'controls' – abounded. As COSHH assessments are now an integral part of risk assessment proper as required under the Management of Health and Safety at Work Regulations ('MHSWR') – they still have to be recorded separately – some of the questions a COSHH assessor should ask are relevant to risk assessment *per se*. They are worth reproducing here because it seems that many lessons learnt during risk assessment under COSHH in the late 1980s have been forgotten.

Firstly, a comparison of risk assessment under COSHH and under the MHSWR, reg. 3:

Under COSHH	Under MHSWR, reg. 3
(1) Identification of chemicals and materials.	Classify work activities into discrete areas.
(2) Seek out relevant information and knowledge on the inventory produced under (1) above.	Look for and identify the hazards.
(3) Assess the risks from above information.	Evaluate risks and decide on adequacy of existing precautions.
(4) Implementation of risk controls.	Record findings and controls implemented.
(5) Monitoring of effectiveness of controls.	Review effectiveness of controls. Revise assessment if necessary.

It will be seen that the five steps in each programme are virtually identical. This means that experience gained from assessing risks from substances hazardous to health under the COSHH Regulations can be put to good use in MSHWR assessments.

Misunderstandings about COSHH hazards arose in a variety of ways but the main problem stemmed from ignorance about the *magnitude* or *amount* of the hazard. Take trichloroethane, for example, a solvent used industrially for degreasing metal components. It is (or was) used in small bottles of white correction fluid in offices. But whereas the solvent was stored in 40 gallon (180 litre) drums or similar in the workplace the quantities in use in offices were in tiny bottles of 20 millilitres. So here was an identical hazard in two different environments. In one, the risk of overexposure was considerable; in the other, it was negligible. But people were putting the same value of risk on both work environments – simply because trichloroethane has an EH40 Occupational Exposure Standard and is toxic and was present!

The overall quantity of a substance hazardous to health used, present, manufactured, or despatched per week, month or year must always be taken into account when assessing risks because these factors, often overlooked, have a direct bearing on the level of risk obtaining. And although the transport of dangerous goods is very strictly regulated the question overall of what form of transport – road, rail, sea, air – is the least hazardous again helps determine the level of risk.

Another area of misunderstanding arose from the *form* the hazard took. Lead, a well-known toxic substance, is a good example. Lead generally comes in three forms – lead oxide powder in large drums for mixing in an industrial process, fumes from soldering using lead/tin alloy and battery manufacture and solid ingots stored in a warehouse. The toxic hazard from lead is present in all three scenarios but only in the first and second cases does it present any real risk to health. Mercury can be mentioned, too. The silver, metallic form of liquid mercury, if ingested, will probably pass through the body's waste disposal system with few ill effects as it is not thought to be absorbed through the gastrointestinal tract. Salts of mercury, however, are a different story and are highly toxic.

FORM, PROPERTIES AND EFFECTS OF THE HAZARD

The assessment of hazards should always make clear what form the hazard takes. Is it a gas, vapour, fume, liquid, dust or a bulk solid? Not everyone to be protected by risk controls will know that nitric acid, for example, is not only notoriously corrosive but is a 'fuming' acid which can destroy the delicate smell sensing membranes of the nose if sniffed.

Physical hazards are often self-evident but electrical hazards can be well hidden from view. Poor or open circuit earth leads will pose risks of electrocution and this type of hazard may not be discovered until someone has been electrocuted. This topic should alert hazard assessors to the hazards and risks created by *poor maintenance* or, as is often the case, no maintenance at all.

A summary of the main properties of the hazard, including its effect where this is not obvious, should always be included in the assessment. Substances assessed under the COSHH Regulations could be corrosive, irritant, highly flammable, narcotic, carcinogenic, teratogenic, mutagenic, an asphyxiant or oxidising. Information intended to reduce or prevent exposure of people to such hazards should explain why they should keep away and should be posted close by the hazard as part of the control arrangements.

CONSULTING WITH EMPLOYEES

Even if the need to consult with employees on health and safety matters in the workplace were not a legal requirement, it is an essential component of both hazard and risk assessment. It has already been mentioned that those working with, or alongside, hazards have a better opinion of the risks posed than does management.

It is also the case that management quite genuinely believe that some opinions of risk offered by workers are exaggerated. This may be so in some instances but they should always be considered and never rejected out of hand. Do not forget that information from manufacturers and suppliers can help you to not only spot hazards but put them in their true perspective if differences of opinion arise between management and employees.

CASE STUDY 4.1

A Manitou lift truck fitted with a JCB-type bucket was working on an agricultural mill site when it became stuck in soft ground. A colleague with a tractor attached a chain to the front of the tractor and the lift truck bucket to try and pull the Manitou onto firm ground. There was a space of about two feet between the two vehicles when the chain was under tension. The lift truck moved three feet or so but the bucket then became wedged against a stack of straw bales.

The two employees then decided to try and extricate the lift truck from the rear so it was necessary to remove the chain to reattach it to the rear of the lift truck. The truck driver got down to uncouple the chain from the truck bucket. As it was under tension the tractor driver shouted to the truck driver that he would move the tractor forward to slacken off the chain. Due to the noise of the tractor and Manitou diesel engines combined the truck driver, who was out of the line of sight of the tractor driver, appeared not to hear the tractor driver's warning to move out of the two foot wide space between the two vehicles. The tractor closed the gap and the Manitou driver's head was crushed between the bucket and the front of the tractor. He died in hospital later.

OTHER TECHNIQUES FOR ASSESSING HAZARDS

So far in this chapter we have been examining the characteristics of hazards rather than risks because it is important to get the hazard in perspective before any assessment of the risk posed is attempted. One technique for identifying hazards and getting a 'feel' for the risk potential involved is Job Safety Analysis or 'JSA'.

Job Safety Analysis involves breaking a job down into its constituent *tasks* and analysing the discrete steps of each task for hazards. Note the important distinction between *job* and *task*. A lift truck driver has a job – that of lift truck driver. But this job will entail a number of quite different tasks, each one possibly involving different hazards. Lift trucks are often used for purposes for which they are neither designed nor intended.

The horrifying tragedy outlined in case study 4.1 was the result of a series of failures to plan the operation by considering the hazards and assessing the numerous risks involved. The overriding need was found to be the resumption of work by both the Manitou and tractor drivers. The tractor was being used with its agricultural implement still attached so the correct towing lug, right behind the driver's seat, was not used. Instead, an improvised connection was made by wrapping the chain around the tractor's front wheel steering members. Amongst other things this made it difficult for the tractor driver to see what was happening in this area – a fact which was to prove fatal.

CHANGE OF SYSTEM OF WORK

Even if the first method adopted to get the lift truck on to firm ground had been properly thought through and an assessment made of the dangers of the operation the abrupt change of *modus operandi* was not considered at all. At this point, a degree of panic would be setting in – stress, in other words. Both men were 'off the job', an expensive piece of plant was stuck in soft ground and slowly sinking further and the preoccupation to get everything 'back to normal' as soon as possible seriously affected judgements. The mill foreman, who had given permission for the use of tractor, had gone back into the mill instead of supervising the operation. Had he been present, in the role of banksman, the fatality probably would not have occurred.

Sudden changes in systems of work to cope with unforeseen developments often hold more perils than people realise. These changes are invariably made 'on the hoof' without any planning or assessments of the dangers inherent in them. The installation of the temporary by-pass pipe at the Flixborough plant in 1974 was a good example of this phenomenon on a massive scale. Such changes invariably create new and different hazards with changes in risk values and must be properly reassessed, not only by law, but from a plain common sense point of view, if accidents such as the one described above are to be prevented.

It is ironic, of course, that a major contributory factor to the fatality was getting production restarted. It was some time before the vehicles could be moved; the police, HM Agricultural Inspectorate, insurance assessors, senior company management, were all involved. Legal matters, post-mortem and attendance at the coroner's court, etc., took up considerable time – as is the case when an employee is killed at work. Production was at a standstill for many days and the total cost to the company concerned was enormous.

JOB SAFETY ANALYSIS AND ABNORMAL TASKS

In the past, the HSE used to mount yearly campaigns to promote safety at work in particular areas where accident statistics were increasing. One of these campaigns was entitled 'Deadly Maintenance' and dealt with the dangers and risks of maintenance activity of all types. The main theme being promoted was that maintenance was often carried out on an emergency basis with management agitating to resume production. Consequently, the maintenance staff were always under pressure to get the job done as quickly as possible and not always as safely as possible.

Maintenance is akin to a departure from laid down safe systems of work which have been drawn up on the basis of proper identification of the hazards and assessment of the risks involved. This forms the basis of JSA and the ensuing safe system of work should embody the controls decided on as a result of the assessment. Whereas routine or planned preventive maintenance is usually drawn up on the basis of safe procedures, emergency or breakdown maintenance is not. Maintenance and other work carried out by contractors is often (but not always) carried out on a breakdown basis.

Where a breakdown maintenance job hits a snag, the often improvised change in procedure often incurs more hazards which, in an atmosphere of increased pressure to complete the job, are overlooked, or dismissed as trivial and not worth worrying about. Case study 4.1 can be used as an analogy to support this contention. The attempted recovery of the lift truck can be compared with a process plant breakdown in that it was a *departure* from the normal safe system of work. The failed

attempt was the snag. The hasty, ill thought out decision to remove the chain, then under tension, to relocate it, was probably not considered a hazard, therefore no risk was seen to exist. As it was, this move (and no pun is intended) proved to be the weakest link in the recovery attempt.

INVESTIGATIONS INTO JSA FAILURES

When investigations are carried out into incidents such as the above, the reasons given for failing to stop and think what one is about to do can be quite revealing. A stock answer, quite well intentioned, is 'There wasn't time; it was an emergency, you see!' or 'We've always done it this way', and an old favourite, 'Nobody told us'. The question of time to carry out a risk assessment is a vexed one as people often think in terms of a ponderous, long-winded, time-consuming, paper chase. Many risk assessments are nothing of the sort. When one decides to cross a busy street a mental risk assessment is performed and a decision made in seconds. The car driving example of risk assessment in chapter 3, above, entails continuous assessment all the time one is in control of the vehicle.

However, while standard, repetitive work such as production line procedures still need risk assessment, the hazards and attendant risks become obvious and generally do not vary too much. Such assessments can start with JSA which should provide all the information necessary to complete the first three steps of the HSE's *Five Steps to Risk Assessment* without any problems.

It is in cases of non-routine work, breakdowns, problem situations and situations where there is a sudden need to vary the system of work that the need to pause and think – is it safe ? – is not just a slogan but a crucially vital step.

CONCLUSIONS

This chapter has opened up the question of risk assessment to examine the wider aspects of real-life risk control. In particular, it has drawn attention to the need to consider the *hazard* quite closely and especially to look at important aspects of the same. Having a well-known hazard on the premises does not mean excessive risks are created or that unnecessary control measures have to be paid for.

The importance of assessing the risks of 'non-standard' or 'abnormal' working is stressed with a real-life tragic case study as an illustration of the need to think before acting. Job Safety Analysis which has been around for a long time is also revisited and a useful comparison with COSHH risk assessments is also made.

Maintenance activities, which have long been a major cause of serious accidents, have been examined in some detail and the difference in risk levels between planned and breakdown maintenance illustrated. The need to quickly review safety measures when unexpected snags arise during an already assessed work procedure and think before proceeding – often on an *ad hoc* or improvised basis – has been shown to be imperative.

CHAPTER 5

Risk Assessment *and* the Preconditions for Failure

Much literature intended to introduce employers and others to risk assessment tends to deal predominantly with 'single line' risks. These are risks emanating from identifiable hazards such as machinery, substances hazardous to health, systems of work such as operating a fork lift truck or other piece of equipment, defective work equipment, storage of flammables, and so on.

There are a great number of hazards creating such risks in many workplaces. They are controlled, accepted, ignored or simply taken for granted. 'We just work round it!' said one employee recently when asked about a defective control on a large item of production plant. These hazards should be dealt with. They may appear to be of a minor nature to those who are over-familiar with them but as the saying goes 'it's the little things that cause the most trouble'. It often is.

PRECONDITIONS

The expression 'preconditions for failure' in the heading of this chapter can be better under-stood from the following explanation. 'Failure' refers to the materialisation of a risk, in this context, and the materialisation is often quite severe – as in the case of the Hixon crossing catastrophe. 'Preconditions' refers to factors that are often unrelated to each other and in their own right can be quite innocuous and apparently free from any hazard or risk. On the other hand, they are clearly a hazard and pose a definite risk but this is not readily apparent. However, these preconditions can come together without being noticed and, at a certain moment a catalyst is introduced and a *much greater* risk materialises.

In the Hixon case study (2.1), the catalyst was the low loader being immobilised on the crossing *at the time the express train was due*. Had the train been earlier, or later, the risk of collision would still have been present but would not have materialised. But the preconditions for that disaster were still present and sooner or later something *would* have become immobilised on that crossing and the collision would have occurred.

23

MULTI-CAUSALITY

The concept of preconditions for failure is often known as the 'multi-causality theory' which refers to the fact that there may be more than one cause to any failure. The theory states that the contributing causes combine together in a *random* manner to result in a failure/loss/disaster/accident. Risk assessment on this basis needs to identify as many of these potential contributory causes as possible. In reality, the reactive approach to failure, which seems endemic to the United Kingdom, usually operates this process in reverse by identifying the preconditions *after* the failure. The Hixon incident, along with scores of others, is proof enough of this.

The way to approach risk assessment of less simple situations is as follows:

(*Note:* the Hixon incident was such a situation but appeared more complex because of the national, as opposed to local, status of the parties involved.)

1. Ask yourself the question: What is the most *unwanted* event that could occur at this site/installation/process/plant, etc.?
2. What are the factors/preconditions/minor events most likely to create this event?
3. How could they come together to create this most unwanted event?
4. What measures exist/are required to prevent their coming together?

FAULT TREE ANALYSIS

This technique can be used for risk assessment where likely contributory causes to a failure can be identified to prevent the failure. It can also be used as a post-event investigative tool to determine how (and why) a failure has occurred. Its main virtue is that it is a *practical* widely used technique which enables root causes of likely failures to be established as well as the causes of failure which have occurred.

The use of fault tree analysis in risk assessment where multi-causality is very likely to be present is both simple and revealingly interesting. It develops the four-point model above, starting with the most unwanted 'top' event. Where a highway intersects a railway at the same level, the most unwanted event must be a collision between a road vehicle or pedestrian and a train.

Secondly, all the proximate or obvious causes (the contributory factors) which could lead to a collision must be identified. The most obvious one in the above situation must be the presence of a fast moving train and a road vehicle at the crossing simultaneously.

Thirdly, the likely reasons that could permit this situation to arise, against all odds, perhaps, must be detected. These are the prime preconditions and must be traced back to establish all conceivable ways in which they could arise. In the Hixon case, an appalling lack of co-operation and information exchange and general widespread ignorance on a number of different subjects prevailed. The Official Inquiry into the disaster found, for example, that the Ministry of Transport's part in the incident 'was not clear' and that 'some of the Ministry's workings in the Hixon affair disappear from view as do connections in an electrical circuit when they encounter a "black box"'.

Fourthly, each precondition thus identified is considered to decide to what extent controls are required to prevent any future contribution to risk materialisation.

By using this system of tracing back during the risk assessment process, the likely causes of failure are more likely to be clearly and objectively identified. Thus effective control action can be taken as appropriate to ensure that the risks do not materialise.

ACCIDENT/INCIDENT INVESTIGATION

At the time of writing, the Health and Safety Commission ('HSC') has issued a consultation document on proposals to introduce a compulsory duty for all companies and organisations to investigate all reportable work-related accidents, ill health or near misses that could have resulted in serious injury. Whether or not this proposal becomes law seems immaterial; all such events should be investigated for the benefit of the financial health of the company, as well as the general health of employees.

Such investigations can utilise the fault tree analysis technique as outlined above. The only difference is that the most unwanted, top event has actually occurred. The investigative process is exactly the same as for risk assessment.

CONCLUSIONS

This chapter has endeavoured to explain how the majority of risks are made up of a number of factors or causes – hence the term 'multi-causality' – and are rarely simple and straightforward. Hazards too can be complex and multi-faceted, often consisting of a number of preconditions which, until they come together, do not constitute hazards as such in their own right but on combining, can be disastrous. There is a human behaviour analogy here. A football supporter on his own, away from the crowd, can be a normal, unassuming individ-

ual. But in a crowd of fellow supporters, during or after a match, he can be quite a different person.

The term 'root cause' has been mentioned in this chapter. This expression is commonly found in accident investigation where the *actual* cause(s) of the event remain hidden and the investigation stops at the obvious or 'proximate' cause. An example would be where a worker has been injured by operating a machine without its guard. The investigation into the accident reveals (admittedly fairly obviously!) that failure to use the guard is the proximate reason for the accident. The remedy? Replace the guard and instruct the worker to use it.

The correct way to investigate this incident is to ask *why* the guard was not being used; to ask why maintenance allowed the machine to be used without the guard; to ascertain why the supervisor permitted the machine to be used and so forth. This line of investigation will almost certainly elicit the root cause(s) of the accident. Now, when assessing risks relating to the use of machinery, one should ask oneself whether there is any history of employees dispensing with guards on machinery and similar equipment or whether there is any *risk* or likelihood of this happening and adopt a similar investigative approach. Risk assessment frequently starts with asking 'What if ... ?' followed by a mental simulation of the most unwanted situation.

Using the guarding example, the risk assessor would be aware that the machine *per se* is a hazard. Its potential for injury is increased considerably if used without its guard. But removal of the guard enables the worker to increase his output. What is the risk, or probability of the worker removing the guard against instructions to keep it in place? If the route recommended for investigation of an accident on the machine is followed the outcome enables a far more effective evaluation of the risk (step 3 in the HSE's *Five Steps to Risk Assessment*) posed by the machine and its operation to be made. This should lead to devising and implementing more effective and pragmatic controls.

CHAPTER 6

Evaluating Risk

The dictionary defines the term 'to evaluate' as 'determining its worth'. All risk management programmes include a step requiring the assessor to 'evaluate' the risk assessed by the identified hazard. Very few give practical guidance on how to carry out this evaluation. The HSE's *Five Steps to Risk Assessment* simply state (see step 3): 'Evaluate the risks and decide whether existing precautions are adequate ...'. There is no worked example of an evaluated risk provided or any real indication of how to go about an evaluation.

To be fair, however, the HSE does have to maintain a degree of impartiality in such fields and one can imagine hordes of businesses asking the HSE to provide them with worked examples of their own risks if they published a specimen of one evaluation from one particular industry. And there is no such thing as a standard evaluation – each and every risk evaluation is specific and individual – personal subjectivity will see to that.

EVALUATION AS APPLIED TO RISKS

In evaluating risk it is not so much the actual *value* of the risk that is being evaluated but the cost to a company of the materialisation of that risk. In other words, 'What if ... ? How much will it cost us if things go wrong?' Risk evaluation is often expressed in *outcomes* in terms of the materialisation of risk and these can be just one or several in number. The outcomes of a serious fire, for example, are likely to be more than just damage to premises. Consideration of *all* possible outcomes leads to more effective control measures.

Careful evaluation of risks also assists the risk assessor to decide more objectively which risks are tolerable and thus acceptable and which are not. This stage includes the cost-benefit analysis input explained in chapter 2, above. Evaluation was mentioned in chapter 4 with particular emphasis on its application to substances which come under the COSHH Regulations. In this field, particularly accurate evaluation helps prevent gross over-expenditure resulting from ignorance, as case study 6.1 shows.

CASE STUDY 6.1

During the compilation of its COSHH inventory, a company finds
it uses sodium hypochlorite in two different situations with
the products supplied by two different suppliers. One source,
used for sterilisation purposes, is in a concentrated form
under the trade name 'Hypo'. The same chemical from another
supplier is used in a far weaker form, which has to be further
diluted before use, and is called 'Clean concentrate'. This
product is used for washing down canteen tables and furniture.

The use of hand, eye and face protection along with apron and
rubber boots, which was appropriate and essential in the 'Hypo'
case was thought to be necessary in the 'Clean concentrate'
case also because 'the same chemical was being used'. All that
was really necessary in the 'Clean concentrate' case were
domestic type lightweight rubber gloves.

PANIC MEASURES

It will be evident that a careful evaluation of a risk is an essential part of the risk management process, as apart from leading to decisions about the correct control measures it can save a company considerable expense. It can rule out the need for costly, unnecessary control measures. To date, there have been numerous cases where just the *presence* of a hazardous material in a work environment has led to what can only be described as panic measures to 'do something about it'. The main misunderstanding seems to centre on the question of exposure.

ASBESTOS

Probably the best example of widespread misunderstanding about a hazard is the official attitude towards asbestos. The very mention of the dreaded word conjures up fear and overreaction because of ignorance about the material. It is worth spending a few moments on this subject.

It is said that asbestos-related diseases are currently killing 3,000 people per year in Britain. But the vast majority of these deaths relate to people who were exposed to asbestos in the 1950s and 1960s when its use was widespread. And the scientific evidence on exactly what levels of exposure cause asbestos-related diseases is unclear – even today after over half a century of research.

Increased publicity recently regarding the possible health risks from asbestos has led to the suggestion that there is a danger to occupants of buildings which incorporate asbestos-based products in their construction. Although there was no evidence to support this suggestion, several independent dust surveys were carried out in the past in hospitals, schools, factories, shops, places of assembly and residences where asbestos products such as asbestos cement sheeting, insulation boards, sprayed asbestos and other asbestos building materials of one type or another had been used in their construction.

The results were extremely detailed and comprehensive. In a nutshell, however, the average mean of *all* readings amounted to 0.012 fibres per cm^3 of air against an official 'allowed' concentration of 0.4 fibres per cm^3 of air for a working lifetime of 50 years.

Asbestos has now been eliminated from brake linings notwithstanding its excellent properties as a reinforcing agent which maintains its strength and stability throughout a wide range of temperatures and pressures. This was despite exhaustive tests by government agencies and laboratories all over the world, who found that over the widest possible range of applications only *trace* amounts of asbestos dust were found to be released into urban atmosphere. These were over a thousand times *lower* than the permitted exposure levels set for occupational purposes.

Asbestosis is a disease of the lungs associated with the inhalation of high concentrations of asbestos fibres over many years. You will not contract the disease by working in premises with asbestos building products where the asbestos is properly sealed, encapsulated or otherwise in good condition. Nor will you get it from drilling a few holes in the wall to put up bookshelves or similar. Slight damage to asbestos lagging, e.g., on a pipe elbow can be repaired. Where asbestos installations are in good condition, you may care to make a note of where it is so that it can be checked from time to time for any damage. *You do not* have to create an 'Asbestos Register' or attend courses on asbestos management or purchase sophisticated 'Asbestos Management' registers. If you have any concerns about asbestos on your premises then HSE's free booklet, *Managing Asbestos in Workplace Buildings* (INDG223), is all you need.

EVALUATING TOLERABLE RISK

Tolerable or acceptable risk has already been mentioned in chapter 2, above, and has been generally defined as the level of risk (also referred to as 'residual risk') which is acceptable to the undertaking concerned. It also means that the risk has been reduced to its lowest level which is reasonably practicable. The factors in the evaluative process to determine tolerable risk include:

- the cost of further managing the risk;
- the benefits of further control measures; and
- the likelihood of materialisation, and consequences of, the reduced level of risk.

This reduced level often emerges after the following options have also been considered:

- *Risk avoidance*
  Where the risk is eliminated altogether. An example would be the substitution of a safer product or process.

- *Risk reduction*
  Where the risk is reduced to acceptable levels by suitable control measures.

- *Risk export or transfer*
  Where a particularly dangerous step in a manufacturing process, for example, is 'farmed out' to a specialist contractor and the product then brought back to the parent company's production line. *Transfer* can mean transferring the risk to insurers.

- *Risk retention*
  Where the risk is fully retained and dealt with by the company concerned.

PROBABILITY

The other factor which has to be considered to carry out a valid risk evaluation is the *probability* or otherwise of the risk materialising and the likelihood of harm occurring. This can be a judgemental process if previous materialisations have been rare or non existent or statistics and past records can give an indication of likely frequency of recurrence.

Much literature suggests a ranking on the following lines:

Probability	Interpretation
Likely/ frequent	occurs frequently, event only to be expected
Probable	not surprising, will occur several times
Possible	could occur sometime, possibly on a random basis
Remote	unlikely to happen but still conceivable
Improbable	very unlikely, probability not nil but fairly close to it

This theoretical approach can be questioned. If a risk materialisation is ranked as 'occurring frequently, event only to be expected' should not some control measure to *prevent* such 'frequent occurrence' have been implemented long ago? However, if a new, first-time situation is being assessed, then it is in order to use such a category.

Conversely, if the probability is ranked as 'very unlikely, probability not nil but fairly close to it', one might question whether the risk so assessed really is a risk at all. Again, it is in order to allocate this category to a risk which is being assessed for the first time or after control measures have been implemented for a risk which was previously higher ranking.

CONSEQUENCES

The consequences of materialisation then have to be considered before any control activity is entered into. Deciding on the consequences is not an exact science as some works on risk assessment would have us believe. If one considers there is a serious threat to one's market via the activities of a competitor then the materialisation of that threat will most certainly result in a loss of profit. If a work environment poses a threat to human life and limb there is no such 'guarantee' that risk materialisation will result in a fatality, major injury, disability and so on. People can be remarkable survivors.

THE WAY FORWARD

Faced, then, with such uncertainties, what should the SME manager do?

The first thing to remember is to keep it simple, as advised in the HSE's *Five Steps to Risk Assessment*. Secondly, avoid charts, graphs, matrixes, nomograms and similar devices. These have the effect of forcing decisions and entries into little labelled boxes. The label is not always appropriate.

Allocate a ranking to the risk from a simple 'high, medium or low' scale after considering the probability of the hazard causing harm or loss.

Check what control measures are already in place. These may be more extensive than you think and you may not need to do anything else. However, do check whether there is a *legal* requirement to provide a specific control. Examples would be adequate vapour extraction from a paint spray booth; personal protective equipment for a variety of situations where exposure to harmful substances is possible; or fail to safety interlocking access points to industrial robot enclosures.

Always bear in mind that *information, instruction, training and supervision* are all control measures and are fundamental techniques in any risk management programme.

If there are any accepted industry standards relating to control of risks on your premises, you should check to see whether you have these in place. Realistically, though, many workplace risks are closely related to the actual work premises and rarely, if ever, are two premises identical where the same processes are being carried on.

The HSE make a very practical suggestion for the SME manager undertaking risk assessment – possibly for the first time – and that is: think for yourself. Think and ask yourself whether you have done all that is reasonably practicable to make your workplace safe. Get a second opinion. It is often very revealing to seek the views of an outsider who is familiar with your operations but is not on your premises on a regular basis. Think in terms of the 'People/Place/Method' model explained in chapter 3, above, and remember that there are these three dimensions to every risk.

CONCLUSIONS

This chapter has concentrated on practical evaluation of risks which threaten the assets of an organisation. It has highlighted quite deliberately ways in which overkill can quite easily occur when ignorance inadvertently allows this or when public and media pressure get out of control. The asbestos scare is typical of the latter situation. A well-known and respected publishing organisation recently issued a broadsheet stating that '3,000 people were dying each year through asbestos-related diseases'. It did *not* state that these people were exposed to asbestos 50 or so years ago - long before the current stringent controls on asbestos were implemented.

The chapter has tried to inject a practical theme to risk assessment. The United Kingdom is not well-versed in this aspect of management as the daily papers prove. Every day we read of failures – in insurance, the financial world, in Government, on the farms, in the world of physical risk, social risk, and so on. At the same time, the chapter has tried to keep the subject simple and down to earth. It has made extensive use of past catastrophic failures as case studies to illustrate how badly things can go wrong – particularly in the context of preconditions for failure. These can exist in even the most modest of SMEs, as well as large corporations.

Throughout this book in general and this chapter in particular, the approach to risk management has been to use the *qualitative* approach to assessment, i.e., the value-based and objective approach, as opposed to the *quantitative* method which is largely mathematics-based, often employing operational research methods. The latter approach is best suited for large, complex, high risk installations such as the petro-chemical industry. For the smaller and more conventional undertakings, the qualitative approach is preferable.

Selection of Appropriate Control Measures

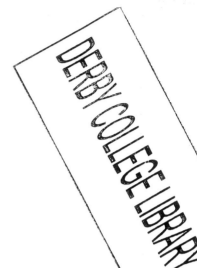

It may seem difficult to believe but many risk assessments stop short once the hazards have been identified and the risks partly evaluated. It was the same when COSHH assessments were first required by law. Employers assembled all the information they could lay their hands on – most of it from suppliers – and did nothing further. The whole point of the assessment exercise was missed and that was to prevent dangerous exposure of their staff to the substances about which they knew practically everything!

Identification and evaluation of risks alone will not reduce risks even if it is argued that people will now be more aware of them. Neither will these two measures alone meet the legal requirement to reduce risks to a level which is reasonably practicable. The selection, implementation and periodic review of appropriate control measures is, therefore, a critically important step in the assessment process.

In considering control measures the following objectives must be met:

- The control must be effective and reduce the risk to tolerable levels, so far as is reasonably practicable. The risk of injury from a band saw, for example, can never be tolerable. Once the limits of mechanical guarding have been reached, training, instruction, supervision, etc., have to take over.
- The control must meet any relevant legal obligations or standards. Where personal protective equipment, for example, is provided (as a *last* resort) it must comply with strict EU standards.
- The control must be realistically affordable in terms of time, resources and cost (see case study 7.1).

CASE STUDY 7.1

A company operated an industrial freezing plant to freeze freshly harvested vegetables as soon as they were picked. The plant operated continuously until the harvest was complete. The plant was then defrosted and during one defrost cycle the welding of a blanked off pipe under pressure gave way. The 4″ diameter blanking disc went through the plant wall like a projectile from a gun with a loud bang. Nobody was injured but an elderly employee died from a heart attack.

The HSE Inspectors decided to require the company to have every weld on all its refrigeration plants up and down the country subjected to an X-ray to detect any patent defects. This order was withdrawn on appeal on the grounds of:

· prohibitive expense; and
· the fact that at the time there were only five people in the country judged to be competent to give an accurate interpretation of any X-ray photographs of such welds.

HIERARCHY OF CONTROL

As in the case of machinery guarding, there is a hierarchy to be followed when selecting control measures. In risk assessment the hierarchy ranges from *elimination* to *information, instruction, training and supervision*. These steps are explained below.

Elimination

The best control measure is to eliminate the problem altogether, i.e., discontinue the process, etc., giving rise to the risk in the first place. It is also likely to be the most unrealistic option. The best time to consider this control is during a design stage or when plant, processes and so on are undergoing major change or modernisation.

Substitution

If a lower order hazard or risk can be substituted for a higher order one without materially increasing costs or sacrificing efficiency then this option should be considered. Substitutes *per se*, however, do not enjoy a terribly good reputation mainly because people are often prejudiced against them on principle. However, there are real benefits in the world of manual handling risks, for example, in reducing the size of a load which requires frequent movement although this is not usually without extra cost.

Engineering controls

These controls are usually expensive but effective and are considered where the risk

CASE STUDY 7.2

Chicken portions are prepared for the retail market at a food processing plant. The prepared chicken carcass is presented to a high speed, exposed cutting device and the carcass initially quartered and further subdivided as required.

Operators engaged in this work frequently suffer cuts, etc., to their hands and fingers (look for the extra finger tip in your next purchase). The company tried all types of hand protection, chain mail, coverall and other gloves, push stick application of the carcasses to the cutter - all of no avail. The operatives stated that they felt more vulnerable with hand protection than without. The HSE agreed and advised the employer to spend extra time on training, which it did. This kept the injury incident rate down to an acceptable (?) level.

can neither be eliminated nor substituted. Noise is an example. Noise reduction measures are both difficult and expensive but can be achieved with acoustic enclosures which isolate the hazard (e.g., compressor rooms), acoustic refuges where people are isolated from the hazard (e.g., in engine test rooms), acoustic baffles, and so on.

Personal protective equipment ('PPE')

PPE should only be considered when the more effective controls have been rejected as not reasonably practicable. PPE is relatively low rated as being unreliable for the following reasons:

- it only protects the user; the risk itself, e.g., noise, is unchanged;
- performance of PPE usually deteriorates with use;
- PPE relies on the correct use and

maintenance by the user; and
- the use of PPE can introduce new hazards for the user.

The last point is particularly significant. *Any* form of clothing, cover or apparel on the body reduces the sensitivity and dexterity of that part of the body, as is clearly evident with the hands when gloves are worn (see case study 7.2).

PREVENTION OR PROTECTION?

Another way of looking at control options is to consider them as either *preventive* or *protective*.

- Vehicle brakes will *prevent* the vehicle crashing; safety belts will *protect* the driver if the brakes fail or the vehicle crashes anyway.

- Local exhaust ventilation will *prevent* the build up of an air contaminant; a respirator will *protect* an employee if the LEV fails.

35

- Correct flammable liquid storage will *prevent* spillage and fire; fire extinguishers will *protect* life and property if used appropriately.

(The above examples are for illustration only; they are, of course, subject to numerous qualifications.)

MONITORING, REVIEWING AND RECORDING

Monitoring

The need for constant monitoring goes back to the 'Place/People/Method' model for assessing risk outlined in chapter 3, above. Everything degrades downwards whether it is dynamic or static and monitoring is essential. Monitoring can range from continuous condition based monitoring, e.g., the fuel gauge on a motor vehicle to a quick check on whether one's television aerial is still in place after a high wind.

For all control programmes an effective monitoring system must be established. All too frequently this vital phase in all kinds of initiatives, projects and other ventures is neglected. Often the excuse is given that funds are not available: 'it's in next year's budget'. As far as risk control measures are concerned, these *must* be monitored regularly for effectiveness if the risk in question is still present.

An effective monitoring programme can be drawn up based on a series of simple questions, the answers to which will determine whether further action is necessary. There is no fixed set of questions; these should be drawn up by each individual undertaking. The following form a sample of such questions:

- *Have procedures for monitoring controls been properly established?*

This may seem a self-evident question but followed by the questions 'what are they?' and 'are they working?' can reveal some startling deficiencies.

- *Are the monitoring frequencies for specific hazards laid down?*

If air contaminants can be present how often are the levels checked?

- *If a Local Exhaust Ventilation System is installed, has a test/examination frequency been established?*

And more to the point, is it adhered to or postponed when production or financial pressures emerge?

- *Are inspections done at the actual site concerned or from a desk?*

The Inquiry into the Littlebrook 'D' power station, Dartford, Kent, hoist failure in 1978 revealed that although the hoist winding rope snapped due to rusting through the maintenance records indicated that the winding rope was inspected and lubricated daily!

- *Is PPE provided and maintained, where necessary?*

In spite of the Regulations on PPE, there is still much uncertainty about the employer's need not only to provide PPE where necessary but to make sure it is properly used and maintained.

- *If reviews and monitoring indicate the need for changes how effectively is this information disseminated?*

A root cause of all failures is the failure to have communicated some vital information. The Hixon disaster mentioned throughout this book forms a catalogue of communication failures.

Recording of assessments, monitoring results and reviews

It is essential that records are kept and maintained in this field. Apart from the efficiency that well-kept records produce, there are financial and legal benefits to gain. These days, the first thing an enforcement officer usually asks to see if an accident has occurred is the victim's training records. The second item will be the risk assessment(s) of the work environment involved.

Well-maintained records can also indicate where excessive maintenance or unnecessary training might have taken place thus contributing to waste of resources. But often more to the point, for an employer defending his position in court, the presence of well-maintained and up-to-date records will help his case considerably.

Do not forget the present day maxim:

'You are unlikely to get into trouble with health and safety law if, despite your best efforts, you get it wrong.

The trouble really starts when you haven't done anything about it!'

CHAPTER 8

Risk *and* the Human Factor

When the Report of the Committee on Health and Safety at Work (the 'Robens Report') was published in 1972, *The Times* newspaper summed up public opinion about workplace health and safety as 'Not exciting, except by accident'. An unfortunate comment and a rather sarcastic play on words by such a supposed responsible newspaper but probably very true at the time. If the Report were published today, one wonders whether the public reaction would be much different. The Report started off by detailing the annual catalogue of industrial deaths, injuries, ill health and disease which were endemic at that time in the United Kingdom. The present campaign about supposed asbestos dangers headlines the fact that 3,000 people a year are dying of asbestos-related diseases but plays down the fact that most of these people were exposed to the mineral over half-a-century ago. (To be fair, the HSE publications state this fact quite clearly.) Yet apart from newsworthy, sensational disasters - such as the explosion at Flixborough in 1974, the King's Cross underground station fire in 1987, the Piper Alpha fire in 1988, the Boeing 737 crash at Kegworth near the East Midlands Airport in 1989, and so on – safety is rarely an important public issue.

PUBLIC INDIFFERENCE TO DANGER

As safety (freedom from injury), and health (freedom from industrial ill health and disease) would seem to be fairly important matters to the individual employee, one wonders what it is in the human make-up that allows, if not encourages, this indifference to personal well-being. This indifference manifests itself in many ways but the outcomes can be devastating. In the examples given above, the root cause and initiating event in each case was human error where commonplace hazards were present and the chances of the risks materialising were high. On each occasion disaster strikes, people in authority, usually government ministers, dutifully pronounce yet again that 'we must learn all the lessons possible from this incident to prevent anything like it happening again'. We never do. And sooner, more often than later, there is a carbon copy repeat. Piper Alpha had many striking similarities to Flixborough, particularly in the field of management failures. And the regular monotonous cycle of failure recommences in spite of the representations from the safety industry.

THE HEALTH AND SAFETY INDUSTRY

Occupational safety and health has become an industry. The European Union has helped in that field. There is a plethora of organisations up and down the country offering training courses of varying length, content and quality. Safety training products – videos, software programs, management games, etc., – are widely available. Universities offer Master of Science degrees in health and safety management. The range of safety protective equipment is huge. The membership of the Institution of Occupational Safety and Health, the premier health and safety organisation in Europe, has rocketed over the last few years. So why does the subject still fail to capture the imagination of the public?

COMMON SENSE

It has to be said that much of the burgeoning health and safety legislation and the widening requirements to make risk assessments are only 'firming up' what we have always in the past considered as 'common sense'. A number of schoolchildren have died on school trips over the last few months. Rules state that teachers in charge must assess the risks of activities on these trips and withdraw the activity if the risk is too great. Obviously, they made faulty assessments in these cases. Yet the HSE states that risk assessment is really only common sense and it does seem difficult to fault this assertion.

The answer to this dilemma probably lies in the fact that what seems to be common sense to one person may seem a complex, difficult situation to comprehend or, alternatively, a situation completely devoid of risk or any form of complication, to another. Common sense appears to have been conspicuously absent in an increasing number of cases if the press is anything to go by. A man tries to drive his car home with a steering wheel locking device still attached because he has lost the key. The fact that the device is designed to *prevent* the car being driven away seems not to register with him until he crashes. Another, on a cross-channel ferry, packed with cars from bow to stern, insists on fixing his steering wheel anti-theft device before leaving the vehicle deck!

SUBJECTIVITY AND OBJECTIVITY

Where people have to make a decision the mental process involved is very complex indeed. Value judgements, prejudices, biases, preferences, experiences and a host of other human characteristics – including ignorance and stupidity – come together to form the actual decision. Sometimes, this happens almost instantaneously; on other occasions, it can take much longer. And just as no two people's fingerprints are identical, no two people's assessments of a risk are the same – even though they may agree on the control measures. One wonders how much store one should place on a risk assessment carried out by the motorist on the ferry?

It is generally recommended that risk assessments are carried out by at least two persons to reduce the possible element of subjectivity in the assessment process. More than two people can be expensive in time and resources and lead to the 'committee' syndrome. One assessor should be the person in the undertaking with responsibility for health and safety; the other a senior manager with the authority to sanction the control measures agreed for each evaluated risk. This arrangement can usually produce a reasonably balanced, pragmatic approach to assessment but is not guaranteed.

CONTROL OF TECHNOLOGY

Some years ago, an authoritative source in the world of science and technology suggested that if risk assessment of possible *system* induced errors was not taught more widely on engineering-based educational courses, there would be a danger of the ability to control burgeoning technology being restricted to a slowly shrinking elite which alone possessed this capability. In the case of the Boeing 737 crash at Kegworth, a system which permits the shut down of the (then) serviceable starboard engine when the port one is unserviceable needs some very careful risk assessment (in this example, the term *system* includes the crew). When one looks at the tremendous wealth of fire technology detail derived from the research into the King's Cross underground station fire, and published by the Institution of Mechanical Engineers at an Institution seminar, one wonders why just a tiny fraction of this effort was not applied in a risk assessment of the danger of a major fire breaking out in an underground station in the first place. But of major concern should be the following passage from a paper given by a distinguished member of the Institution of Mechanical Engineers at the above seminar:

'Due to a history of fires on these escalators (the type installed at King's Cross) they were fitted with what is called water fog equipment, which are water sprays underneath the escalator which are manually operated by valves on a landing inside the access door to the upper (escalator) machine room from the tube lines ticket hall.'

These valves were not operated during the fire. No attempt was made by anyone to turn on the water fog equipment. The official Inquiry Report suggests that station staff had forgotten they were there and in any case had neither received adequate training nor had management drawn up an adequate fire drill plan.

The statement quoted above is not developed by the speaker and exists solely in the context of a general description accompanying a diagram of the escalator machine rooms. Other parts of the Report do deal with management failure but only in a general, scatter gun manner.

The words of the Bishop of Grimsby spoken while officiating at the service for those killed in the Flixborough explosion are worthy of note:

'As mankind breaks through more and more barriers of technology with the intention of improving the quality of life, he does in fact make life more precarious.

One slip ... and the abyss'.

It would seem that the greatest threat to health and safety and risk control at work is not posed by machinery, equipment or substances hazardous to health but by human factors – those supposedly accountable for the technology under their control. But when one reads, time and time again, of organisational mistakes which are attributed to 'computer error' when they are computer *operator* or *programmer* error, one may wonder whether much of the UK's management has already been left behind by the most elementary technical equipment. The Fennell Report into the King's Cross disaster found that; 'senior London Underground management made fundamental errors in their approach to safety' and that 'its chairman at the time did not think that safety matters were the strict responsibility of London Regional Transport'.

It also found that the Railway Inspectorate 'misunderstood its responsibilities under the Health and Safety at Work, etc., Act 1974'. The Act had been in force for 12 years and a major inspectorate did not understand its responsibilities after all that time!

DISMISSIVE ATTITUDES TOWARDS SAFETY AND RISK

It seems that there could be a number of reasons for these almost incomprehensible attitudes displayed at times towards occupational health, safety and risk. Just who, one may ask, did the above-mentioned chairman think *was* responsible for safety? and what was the Railway Inspectorate doing to clarify its responsibilities under the Act for over a decade? The answer appears to be – nothing.

There can be a number of reasons for such dismissive, offhand and often trivialising attitude towards safety and risk. Firstly, these topics are often treated with some degree of levity – sometimes deliberately and sometimes unintentionally. What is the reader to make of the following report of last year's opening event for the 'European Week for Safety and Health' due to commence later this year?

'An undercover team of experts carried out on-the-spot checks on police stations throughout South Wales during last year's week. They were making sure that no health and safety laws were being broken. Anyone found guilty was locked in a cell and made to read the regulations before parole was considered.'

The police were presented with a health and safety award by a distinguished national newscaster after this event.

Or the following report in a national daily newspaper:

'A two-year-old boy had been given a toy plastic golf set as a present and took it into the local park, again in Wales, to play, accompanied by his grandfather. Swinging his hollow, plastic club enthusiastically the maximum distance he could propel the thin, hollow plastic ball was 7'. Before he could "play" another shot a park attendant rushed forward to tell the grandfather that his grandson "was a danger to passers-by". The attendant told them to leave the park with an admonishment "You can't play golf here!". A council spokesman confirmed that the park rules stated that golf must not be played in its parks and said the attendant "was only doing his job".'

The cases of the perils of conker gathering and life threatening window boxes have already been described in chapter 1, above.

Examples of banal stupidity such as the above do not enhance the status of what is a very serious subject – more so when you become a victim. To be fair, one can presume that the council, like many others, is terrified of being sued by some ambulance chaser for damages for 'injuries' sustained by the chance event of being hit by the featherweight ball. But who awards the huge, nonsensical settlements which greatly offend the majority of the public? Human factors, that is who.

Secondly, many people ostensibly in senior positions in undertakings which suffer disaster have an engineering background which, surprisingly for some, does not always sit happily alongside health, safety and risk management. University engineering courses are so crammed with technical subjects that little or no time is left for any subjects about people and that includes health and safety at work. In any case the engineer dislikes the human factor, preferring instead to deal with the hard-edged, predictable, measurable and reliable hardware of engineering. When the human factor *is* incorporated into a system, it is regarded as being as reliable and predictable as all the other components, when in actual fact the very opposite is likely to be more accurate.

Thirdly, the designer, engineer or user concerned with the performance of an article, product or system tends to halt his endeavours once that project performs according to specification and is reliable. He does not ask what could happen when something fails in terms of the consequences and the effects on *people* as well as the immediate surroundings. He would probably be oblivious to Judge Du Parcq's ruling on the test whether a part of a machine is dangerous or not as 'whether it might be a reasonably foreseeable cause of injury to anybody acting in a way which a human being may be reasonably expected to act in circumstances which may be reasonably expected to occur'. This 1937 ruling has seen much subsequent legislation covering defective equipment, consumer protection, etc., but the need to foresee the use to which a product *may* be put – for which it was neither designed nor intended – still needs to be borne in mind by design engineers.

Fourthly, the number of slogans, weary safety posters and the amount of safety propaganda has long reached saturation point. As a nation, we seem to be hag-ridden by restrictive legislation every way we turn. Risk-taking itself is gradually being discouraged or even forbidden by law. Too many restrictions on an individual's activities in the name of safety are simply counter-productive.

BOREDOM BREEDS RISK TAKING

Workers engaged on production lines carrying out boring, repetitive work often seek to inject some excitement into their tasks by ignoring common sense safety rules, as case study 8.1 shows.

This installation was judged to be safe and acceptable by the Factory Inspector on condition that operatives' hands and arms were restricted to working only on the product. The line was fairly slow moving and the danger posed by the shackles and stanchions was obvious. There was no pressure on the operatives; the line speed had been agreed by union and management. But boredom and its associated risks had been overlooked and consequently did not feature in the risk assessment of the process.

HEALTH AND SAFETY OFFICIALS

Students on management studies courses are usually introduced to the subject of role conflict within organisations in the first few weeks of the course. This conflict can arise between employees often known as 'company men' who are generalists and whose prime loyalty is to the company and the specialist professionals whose prime loyalty is to their profession. The former improve their position by promotion within the organisation; the latter move from company to company as they develop their professional expertise.

Conflict

Conflict arises when the professional tries to impose change, for example, based on his professional knowledge, which may be objected to by the company manager. The latter tends to look at the wider situation – from the

CASE STUDY 8.1

Workers on a food processing line sat either side of long tables with the products moving along the centre of the table at shoulder height. The products were hooked onto heavy shackles attached to a moving chain. The chain was supported by a long, horizontal gantry which was in turn supported by vertical stanchions bolted to the centre of the table at approximately 12' intervals. The machinery was driven by electric motors.

Due to the danger of getting a hand or arm jammed between the shackles workers were under strict orders not to pass articles through the line to colleagues on the opposite side of the table. Because the processing work was so boring the operatives ignored this rule. They also played a game of 'chicken' whilst the line was running empty by passing their arms through the shackles and withdrawing them at the last moment before the shackles passed through the upright stanchions. Small wagers were won by those withdrawing last.

One young operative miscalculated and her arm was drawn into the upright stanchion wrapped around the shackle in a 'U' shape causing multiple bone fractures. The line was stopped by an operator yanking at the emergency stop pull cord switch but the girl's arm could not be released until an electrician was found to reverse the polarity, and thus the direction of rotation, of the drive motor.

company's point of view – and may feel that the professional is perhaps going a little 'over the top'. This certainly was the case with company 'safety officers' in the early days of the Health and Safety at Work, etc., Act 1974. With very few Regulations, Approved Codes of Practice and Guidance Notes to clarify the general requirements of the Act, interpretations of 'risk of injury or ill health' and 'so far as is reasonably practicable' were rather too severe for some managing directors' liking.

Safety staff managed to alienate themselves from almost everybody and tended to be involved only when absolutely necessary, for example when an accident had occurred or an enforcement officer visited.

Risk assessment in the 1970s

It is not generally realised that risk assessment was, and always has been, the cornerstone of the Act. It marked the sea change in philosophy between pre-1974 legislation and the post-1974 approach. Only comparatively recently have safety advisers and risk managers realised the need to integrate their activities with the corporate day-to-day activities of boards of management to get themselves heard. All too often in the past, the risk manager/safety adviser has been a voice in the wilderness because of the adoption of too headstrong an attitude.

Image and status

Older readers will recall the field day media cartoonists had during the Second World War with experiences of Whitehall bureaucrats visiting industry dressed in (then) regulation bowler hats, pinstriped trousers, polished shoes, rolled umbrellas and briefcases. The late 'Giles' frequently made fun of Ministry of Agriculture employees visiting farms to make various inspections and being led up the garden path through the crew yard, over the manure heap, into the pig sties, etc., to make their inspections. The fact was that these 'Men from the Ministry' were not welcome, were often unsure of their task and were heartily resented by farmers as 'interfering snoopers'. They had a very low status in the agricultural industry.

It is most unfortunate, then, that the following incident was reported in a national newspaper at the height of the tragic foot and mouth epidemic:

'At the crack of dawn ... a Cumbrian farmer resigned himself to the slaughter of his several thousand healthy sheep. Eight slaughtermen in white garb were poised for action when a Ford Mondeo arrived to spill out a young suited official with appropriate demeanour.

He was from the Health and Safety Executive, and announced that before the cull could begin he had to check that there were adequate toilet facilities for the slaughtermen. It is not recorded what the farmer or the slaughtermen murmured under their breath. Lord Rix may be delighted to know that the spirit of Whitehall farce is alive and well.'

The newspaper in question has a circulation of several million readers who will have seen the above news item which pokes fun at the HSE. This is unfortunate, to say the least. For several years now, both the Commission and the Executive have built up a reputation for quiet efficiency and a reasoned approach to regulation of the workplace. The latter's information provision facilities are second to none and its publications are excellent. It has advanced the cause of health and safety at work quietly and progressively with positive results. This one incident, where a complete lack of empathy with the people on that Cumbrian farm was displayed by an official, will not have helped the cause of safety. What degree of risk was involved in this situation?

EXECUTIVE VERSUS ADVISORY ROLE

The argument as to whether a safety adviser or risk manager should have executive authority to 'stop the job' has waxed and waned for many years. A lot depends on the type of industry and the status of the safety person. In the old UK mining industry a safety officer was appointed, often a long-serving employee who may have been injured during years of underground work, and a safety engineer. The latter had executive authority as a senior official, whereas the safety officer usually had an advisory role – backed up, of course, by his acknowledged years of experience, an intimate knowledge of the industry and the respect in which he was held.

The general consensus of opinion is that the safety adviser/risk manager or assessor should not have executive authority and should command an advisory function only. The main justification for this is that managers should retain full authority for all matters in their areas of responsibility including safety. Any suggestion that the safety adviser had authority in risk assessment, for example, could lead to abdication of safety and risk responsibilities by the manager. It is up to the manager to act on safety and risk advice proffered and to be accountable for its implementation. However, this should be done after mutual discussion to agree the 'reasonably practicable' factor in terms of the nature of the risk, the benefits likely to result and the costs involved.

TRADE UNIONS AND FATALISM

Before trade union power waned to its present level, health and safety was frequently used as a bargaining counter against management. It was noticeable that when times were hard and unemployment was growing, health and safety activity was usually at a low ebb. Once the economy improved, activity increased. This is a worrying phenomenon. Workplace hazards and risks do not reduce when times are hard or return to their steady state when things improve. Yet people seem to *believe* that this is the case and nobody appears to be concerned about it. Workplace risk assessment evaluations tended to fluctuate more between the high, medium and low categorisations in sympathy with the economy. Trade unions do back pedal on safety when redundancies threaten. Company directors too, have been known to use health and safety matters to influence decisions at board meeting and score points over fellow directors. Such antics are not designed to get health and safety and risk management treated seriously by those who have the power and authority to make this happen.

Secondly we have, as a nation, a fatalistic attitude towards our own destinies. At work, we have the feeling that 'it' will not happen to us – it is always the other guy. If 'it' does happen to us, it seems to have been ordained, anyway. In the United Kingdom's heavy industrial past, acceptance of injuries and ill health at work was normal. Workers in the Sheffield cutlery industry – particularly the basement knife grinders – had a life expectancy of 35-45 years before 'the dust' (silica dust from the grinding operations) wrecked their lungs. If a youth working in a heavy engineering works complained about noise induced hearing loss, he would be scolded by his mother. 'Your father and your grandfather were all deaf at your age. What do you expect? It goes with the job.'

INDIVIDUAL ATTITUDES

An elderly worker in a Midlands metal fabrication factory told the author how he had long got used to the noise, how much compensation he had received already and how much the union was expecting to obtain for him before he retired. Some employees seem to wear their work-based afflictions with pride. Others have some curious ideas about personal immunity from ill health. A worker at a provender mill in Wales was a member of a rare breed who chewed tobacco. He worked in an area where the average dust levels were over the then maximum permitted level of 15 milligrams per m^3. He refused point blank to use any form of respiratory protection and even the HM Senior Agricultural Inspector of the HSE could get nowhere with him. He maintained that the tobacco juice in his mouth absorbed all the dust he was likely to ingest and expectoration would finally dispose of it. He had no answer to the question of what happened to the dust inhaled through his nose.

RHETORIC AND REALITY

The slogan 'safety before production' is an example of much idealist rhetoric (safe production is far more realistic) which still persists in parts of industry. Where the primacy of production is deeply ingrained in the thinking of managers – who do claim to care about safety – some of their actions appear quite contradictory. It is not uncommon for such a manager to attend a joint consultation meeting and declare that the order of the meeting will be 'production first, then safety, if there's time'. At such a meeting the question of progress on risk assessments might be raised. If it is, the answer will usually be 'noted' and often no further discussion ensues!

One of the most famous examples of rhetoric concerns the collapse of one corner of the Ronan Point tower block in Newham, London's East End, in 1968. The consortium that built it comprised a building company, a manufacturer of prefabricated panels (which were assembled on site into the tower block) and a consultancy (which was also retained by the client – Newham Borough Council). The various parties involved are reminiscent of those involved in the Hixon level crossing catastrophe. In the Ronan Point case, no one realised that the removal of any one vertical load bearing panel would leave everything above it unsupported. A domestic gas explosion on the 16th floor blew out an external wall panel, and instantly the entire eastern corner of the building collapsed like a pack of cards, sending thousands of tons of masonry, along with furniture, plumbing, fittings etc., crashing to

the ground. Miraculously, only four people died and 17 were injured. Nobody had assessed the risk of the likelihood of the failure of a vertical panel.

One of the photographs of the disaster shows the building consortium's site offices in the background, clearly identifiable by its tug-of-war logo, with a large sign over the doorway with the words: 'SAFETY COMES FIRST'.

Rhetorical statements abound in the safety world. Safety policies are notorious for them, containing time and time again written undertakings to 'ensure' and 'provide' to an extent that would bankrupt a company overnight if implemented. It is almost impossible to find a safety policy prepared in the realistic manner in which a cash flow forecast, or similar corporate document is prepared where the risks involved are sensibly and accurately assessed. Anybody – politicians especially – who expresses concern about safety is almost certainly assured of a sympathetic audience because it is unacceptable to suggest publicly that even modest risks should be accepted.

It seems ironic that most car advertisements these days offer safety as a major selling point, e.g., twin airbags, ABS, safety bars inside body panels (impact protection systems). They also offer danger simultaneously in the form of lightning acceleration and three figure top speeds! The rhetoric is safety; the reality is danger!

It is a well-known fact that people are often reluctant to let risk control measures work. Improve a so-called 'accident blackspot', e.g., straighten a sharp bend in a road, and drivers will increase their speed through it. Pass a law requiring cars to have better brakes and motorists will see this as a justification for driving faster.

AVOIDING ACCOUNTABILITY

A cartoon depicting that loveable canine character, Snoopy, exists showing him as captain of a cricket team which has lost a match. His facial expression is quite sly and the wording in the balloon is a play on the words of a past famous sports quotation. It goes something like this: 'It does not matter if you lose the game, what matters is how you place the blame'.

It does seem that denial of responsibility for failures is more common than ever these days even when faced with incontrovertible evidence. The second largest motor vehicle company in the world, after years of denial, has agreed to settle a lawsuit concerning the safety of millions of its vehicles. Some years ago, it was in a similar position denying that the petrol tank fitted in the boot of one of its models was a fire risk in the event of a rear end collision.

The errors of individuals are seemingly denied, condoned or ignored by employers and by other parties who appear to bury individual accountability by misplaced sympathy, a careful choice of words, or, when in authority, an apparently flawed investigation as time goes by.

Thirteen years after an event which has seriously affected the health of some 100,000 or more people, adults and children (see case study 8.2), the Government is to allow a 'definitive review' of the effects of the disaster on the health of those affected.

CASE STUDY 8.2

In 1988, a tanker driver discharged 20 tons of aluminium sulphate into the mains water supply at the Lowermoor treatment works in Camelford, Cornwall. He actually connected his discharge hose to the water supply instead of the aluminium sulphate holding tank. Quite how he could make such a mistake seems beyond comprehension (why were the connections not designed to make a wrong connection physically impossible?). Over 20,000 homes were affected and after the contamination residents complained of a wide range of illnesses. However, a government health advisory panel led by Professor Dame Barbara Clayton in 1989 said there should be no long-term ill effects. Lobbying for a fresh investigation went on for ten years and when in 1999 the British Medical Journal published a report which concluded that victims had suffered 'considerable damage' to their brain function, Dame Clayton's findings began (not unexpectedly) to be questioned. At the same time, it was considered that the high levels of aluminium in the water could be associated with increased rates of Alzheimer's disease.

RISK ON THE RAILWAYS

Rail disasters occupy much headline news these days and are reported as if they are a recent phenomenon when in fact they go back many years. In October 1952, a commuter train at Harrow and Wealdstone station was hit in the rear by the Perth / London express. Minutes later, a two-locomotive Manchester bound train piled into the first smash. Very little is said about human signalling error in the report on the disaster.

Accusations about poorly maintained railway tracks have been flying around quite a lot of late – again as if this is only a recent phenomenon. In 1967, at Hither Green in British Rail's Southern Region a fractured section of track derailed the Hastings / Charing Cross express, killing 49 people. In this case, an Inspector at Hither Green had advised British Rail that he had insufficient staff to look after the lines properly. British Rail admitted at the Public Inquiry that because of its inefficient administration nothing was done about the Inspector's report. Express trains were allowed to continue passing over tracks nearing breakdown point. The rail in question fractured at a joint creating a $5\frac{1}{2}$ inch gap which caused the derailment.

Over three decades later, we have the carbon copy Hatfield crash. What happened, one wonders, to all the lessons we should have learned?

The Paddington rail crash was caused by the driver of one of the trains involved passing a signal at danger. It was not caused by anything else. But one reads that the driver error known as 'signals passed at danger' is now an accepted occurrence by the HSE, the Chief Inspector of Railways and, presumably, the travelling public! The practice has earned

itself such a degree of respectability that it has its own acronym – SPAD. Incidents are known as 'SPADS' and the HSE and Railway Inspectorate dutifully publish statistics relating to these incidents which are quite numerous.

The incidence of SPADS actually *rose* in the wake of the Paddington crash despite improved training, monitoring and refresher courses. Instead of making drivers more accountable for their errors, rail executives say they 'do not know' what caused the increase. Instead, the rail industry, already famous for ingenious excuses for failure, thinks 'the warm weather may have led drivers to dehydrate slightly and become fatigued'. How then, should the intending passenger assess the risk of taking a train journey?

It appears that not only do train drivers drive past red signals; they also ignore train protection warning systems installed in their cabins. The Government is proposing installing costly *automatic* systems which will stop the train if a red signal is passed. One wonders just what responsibility drivers *are* prepared to accept as the risks of train driving are, apparently, progressively abdicated.

Still on the subject of driving public transport, one regularly reads of bus drivers driving double-decker buses under single-decker bridges and removing the top of the bus in the process. The cry goes up to 'remove the dangerous bridge' – which may have been there for decades – instead of requiring the driver to account for his negligence.

REPETITION

The armed forces use a category of training called 'overtraining' where the constant, repetitive information provision is intended to enable the trainee to react instinctively when faced with an immediate emergency. Certain aspects of jet fighter pilot training includes such an approach.

'Overtraining' or constant repetition of safety rules does not work altogether satisfactorily in industry – or elsewhere, for that matter. How many of us pay close attention to flight cabin crews explaining ditching evacuation procedures before a flight? The risk of ditching into the sea is low but the risks of injury or drowning if ditching *does* occur can be very great.

Some years ago, a woman cartridge examiner was 'shot' by the machine she operated at the Royal Ordnance Factory at Radway Green, Cheshire, five minutes after hearing the safety rules read out. The coroner was told that women workers at the factory completely disregarded safety instructions, especially when clearing jams in the machines which checked the dimensions of 7.62mm rounds of ammunition. These malfunctions were only to be cleared by experienced machine setters who alone had the keys to open the breeches. The jams occurred when an incoming round entered the measuring gauge before the existing one had exited the breech. The woman operated the manual control to force the incoming round to push the jammed round out of the breech. Unfortunately, she operated it too hard and the pointed bullet of the incoming round detonated the jammed round acting like a firing pin. The jammed round fired, its bullet hitting the woman in the chest. She died in hospital.

Here is a case where a high-risk activity had been accurately risk assessed and control measures implemented. These included the daily reading out of the safety rules. The dead worker – and her colleagues – virtually ignored these as often as they were read out. Was there a missed risk in the assessment process created by the apathy caused by repetitive reminders of the safety rules?

CONCLUSIONS

This chapter has taken a critical, and somewhat contentious at times, look at the part the human factor plays in assessing and controlling – or attempting to control – risk in the workplace and in social activities. The chapter has been illustrated by a number of anecdotal case studies of incidents which are all in the public domain. The section headed 'The Control of Technology' makes the point that we seem to be increasingly unable to control even conventional technology as time passes. The section headed 'Avoiding accountability', above, introduces a very worrying development. Published examples from the railway industry are used to illustrate how risks are still being taken with passengers' lives as train drivers' inability to act on red signals is condoned by their employers.

It seems that in many ways the human factor has not moved very far from the position identified by the Robens Report in 1972 wherein it is concluded that the biggest single cause of accidents is apathy. Is this apathy a result of *too much* safety regulation, one wonders. In particular, the need to assess the risk in every aspect of work activity may be making employees blasé to the extent that they feel that any risk they encounter will be dealt with by the employer. Like the train drivers.

Index